Ricardo Stedile Neto
Sandra Ana Bolfe

The socioeconomic structure of low-income housing in Santa Maria/RS

Ricardo Stedile Neto
Sandra Ana Bolfe

The socioeconomic structure of low-income housing in Santa Maria/RS

Social plots

ScienciaScripts

Imprint

Any brand names and product names mentioned in this book are subject to trademark, brand or patent protection and are trademarks or registered trademarks of their respective holders. The use of brand names, product names, common names, trade names, product descriptions etc. even without a particular marking in this work is in no way to be construed to mean that such names may be regarded as unrestricted in respect of trademark and brand protection legislation and could thus be used by anyone.

Cover image: www.ingimage.com

This book is a translation from the original published under ISBN 978-613-9-67233-2.

Publisher:
Sciencia Scripts
is a trademark of
Dodo Books Indian Ocean Ltd. and OmniScriptum S.R.L publishing group

120 High Road, East Finchley, London, N2 9ED, United Kingdom
Str. Armeneasca 28/1, office 1, Chisinau MD-2012, Republic of Moldova, Europe
Printed at: see last page
ISBN: 978-620-8-15590-2

SUMMARY

ACKNOWLEDGMENTS

First of all, I thank God for guiding me through this journey, always with the health and courage to overcome the difficulties that have been imposed on me; to the Federal University of Santa Maria, for its quality public education;

To the teachers and staff of the Department of Geosciences, who helped in some way to shape my academic and personal life;

I would like to thank my advisor, Prof.ª . Dr.ª . Sandra Ana Bolfe, for dedicating her time and trust to passing on her knowledge, which went beyond academic life;

To the Santa Maria City Hall and the Municipal Department of Social Development for their help in carrying out the work;

To the residents who were interviewed during the fieldwork;

To my colleagues and friends from the Geography course, with special affection for Jaqueline Barreto, Ligian Gomes and Elisa Araùjo, for their countless words of comfort, as well as the academic contribution they made to me;

To my grandmothers, Glaci Corrêa da Silva (in memorian) and Rosa dos Santos Stedile, who always supported me and encouraged me to pursue my dreams, and were always an example of life and perseverance;

I would like to thank my parents Marco Antônio Corrêa da Silva and Andréa Stedile for always respecting my decisions and helping me to achieve all my goals;

To my uncles, godfathers and godmothers, especially Fernanda Stedile and Glani Medeiros;

And to all the people who helped me in some way and contributed to the construction of this work.

THANK YOU VERY MUCH!

1 INTRODUCTION

The urban space is increasingly dynamic and the science of geography is of great importance in man's relationship with the environment, especially in medium-sized and large cities. In addition, geography is an extremely important factor in analyzing, understanding and seeking solutions to the prospects and obstacles facing the growth of population occupation, as well as the increase in cities, exclusionary technological advances, among others. This is caused almost exclusively by the insertion of capital in contemporary cities.

The municipality of Santa Maria/RS is not far from this reality, and in recent decades has undergone a series of physical and social changes in its urban structure, caused by real estate pressure, which overprices housing, thus generating segregation and a housing deficit.

Santa Maria was divided into two stages of urban structure. The first, between the beginning of the 20th century and the mid-1980s, was focused on the railroad, which put the municipality in the national spotlight as it was located in the central part of the state of Rio Grande do Sul and where most of the passenger and freight trains had to pass. In addition to the economic sector, the town's urban setting was largely influenced by the era of the Maria Fumaça, such as the construction of Vila Belga, which was built for the workers at the railway station, a village that is still one of the town's main tourist attractions today. The influence of the "train era" can be seen in the countless tales, stories and songs written by artists from Santa Maria, in particular the musician Beto Pires, who made his song the unofficial anthem of Santa Maria .[1]

The second stage was marked mainly by the establishment of the Federal University of Santa Maria (UFSM) in the 1960s. With the start of the educational institution's activities, the town became a university hub, being nicknamed the "University City" because it is home to numerous higher education institutions

---

1 See the lyrics of Beto Pires' song in Appendix B.

and has a large commuter population[2] , who live in the city during their undergraduate years.

In addition, Santa Maria has become a national reference in defense, as it has the second largest military arsenal in the country.

Due to its importance in education and defense, the municipality's economy mainly revolves around the civil service. Commerce is responsible for a large part of the jobs on offer in the municipality.

Because of these factors, the territorial unit under study is a center of interest for real estate pressure, which justifies the large number of building projects and gated communities that are currently underway in the municipality.

Population densification, in addition to the real estate pressure in the world's major metropolises and cities, causes numerous structural and social problems. One of the most common is the appropriation of unsuitable areas for housing, usually by the low-income population. These areas are usually on the outskirts of cities, in public or private domain, as well as risk areas[3] , endangering the local population.

Based on this, we can see that the occupied areas often need to be expropriated for public works or repossession. Faced with this prospect, the government has been trying, through the implementation of housing policies, to remove this population from those locations that are considered inappropriate. In order to achieve the aim of these policies, in recent years there has been an increase in the number of popular free housing condominiums, which are intended for these families who are displaced from these unsuitable areas after being contemplated in these housing policies.

In this context, the municipality of Santa Maria, located in the Santa Maria

---

2 People who leave their city to work or study for a period of time.
3 According to the Federal District's Undersecretariat for Public and Social Order (SEOPS-DF), risk areas are defined as areas unsuitable for housing due to the fragility or instability of the terrain caused by nature or the actions of man.

micro-region, was selected as the study site. The main concern of the work was to analyze the social and spatial transformations brought about by the implementation of the condominiums under the federal government's Minha Casa Minha Vida social housing program, and to draw up a map showing the housing developments within the municipality. The map is intended to identify in which area of the city the housing estates are located and why the city council chose these areas.

The municipality of Santa Maria is located in the central region of the state of Rio Grande do Sul (MAP 1), between the slopes of the Serra Geral and the Depressão Central Gaùcha.

With this in mind, the general objective of this research is to identify and analyze the popular condominiums of the Minha Casa Minha Vida program in the municipality of Santa Maria/RS, in order to understand/explain the actions of public policies in relation to the basic needs of its low-income population. Specifically, we sought to: (1) analyze the process of selecting and contemplating affordable housing; (2) map the MCMV condominiums in the city of Santa Maria/RS; (3) carry out a socio-economic survey of resident families; (4) verify the current conditions of residents in relation to the physical and social infrastructure of residential condominiums.

Map 1: Location of the Municipality of Santa Maria/RS.

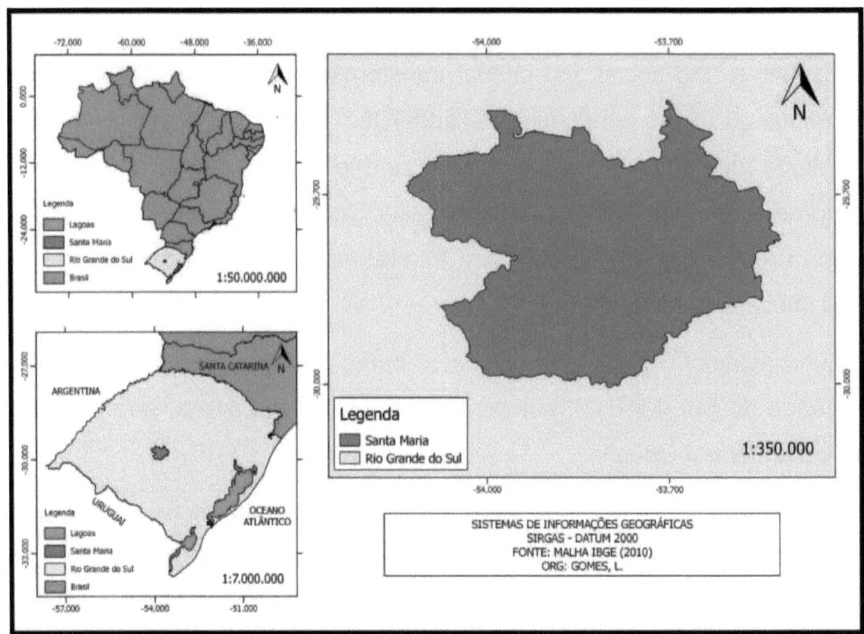

Methodologically, the research was carried out in stages. Firstly, the concepts that guided the work were consolidated, based on a vast bibliographic reference. The conceptual framework structured the theoretical-methodological reference of the work, through specific bibliographies on the subject under study. At the same time, a historical survey of the municipality was carried out, with emphasis on the urban structure of the city, as well as the economy and the population occupation process.

In the second phase of the research, information was sought from specialized bodies on the subject of study, such as the Brazilian Institute of Geography and Statistics (IBGE), the Santa Maria City Hall in conjunction with the Housing Secretariat and the Municipal and Social Development Secretariat.

The third stage of the research was based specifically on fieldwork, where the residents were visited and a questionnaire (APPENDIX A) was applied in order to carry out a socio-economic survey of each housing complex. The target audience for the interviews were the residents covered by the MCMV program.

These questionnaires were applied over five days of fieldwork, three days in the Zilda Arns and Dom Ivo Lorscheiter housing estates, with 25 and 28 interviews respectively, and another two days in the Videiras housing estate, where 21 questionnaires were applied. Initially, it was hoped to interview 10% of the residents of each housing estate, but during the fieldwork the generalization of the data was noticed, so it wasn't necessary to interview 100% of the intended sample. Therefore, 5% of all residents were interviewed. By carrying out the fieldwork, it was possible to insert oneself into the reality in which these beneficiaries are inserted, thus enriching the analysis of the results.

The final stage of the work consisted of analyzing and consolidating the results obtained, plus the preparation of maps spatializing the residences. The software QGis 2.10 was used to create the maps. The Google Earth satellite image tool was used to extract the base image for the maps. The construction of these maps is justified by the observation that the municipality does not have these housing estates spatialized. In the future, the aim is to make the maps available to the Housing Department so that they can be included in its archives.

When all these steps were completed, the final version of the results and final considerations were written.

2 THEORETICAL BACKGROUND

This chapter seeks to understand the concepts that guided this research, as well as understanding the geographical space. It recalls the process of urbanization and industrialization, as well as the concepts of popular housing and housing policies.

2.1 Urbanization and Industrialization

Over the last century, Brazil has undergone a series of changes in the structure of its cities, such as the rural exodus, which intensified during the 1960s with the policies of President Juscelino Kubitschek, who made it possible for international industries to come to Brazil, mainly in the Southeast. This factor transformed Brazil, which had previously been an agrarian country, into an almost entirely urban country. This fact is proven by Fedozzi, who says that "the population of cities increased by more than 60 million people; 29 million during the 1980s alone, with urban areas now concentrating two thirds of the population" (FEDOZZI, 1999, p.17).

This migration process in Brazil was driven by the process of "conservative modernization" that intensified during the 1964 military dictatorship. This model of modernization has some important characteristics, such as:

A process of concentration of wealth, urban land, selective access to public facilities and services, unprecedented in the country, making the centers of capitalist development in Brazil emblematic of the situation of inequality, urban segregation and environmental degradation. (FEDOZZI, 1999, p. 17)

With this process of accelerated urbanization in the big cities, a large part of the new population that began to arrive in the urban centers, without having the resources to live in the regularized neighbourhoods, began to occupy unsuitable areas, in irregular housing. This occupation led to the creation of Brazil's large favelas, which are known the world over. According to Corrêa (2004, p. 9), urban space is "[...] fragmented and articulated, a reflection and social conditioning factor, a set of symbols and a field of struggle".

In addition to the housing problems, this misguided occupation of public or private areas in an inappropriate manner has generated numerous social problems, such as education, health, sanitation and public safety.

According to Corrêa (2004), society also has the power to condition the space in which they live, through the social forms established by the population. This conditioning takes place through social agents, who produce and consume space through the accumulation of capital. These social agents are: owners of the means of production, landowners, real estate developers, the state and excluded social groups.

Industrialization is another factor that has led to a great demand for housing in Brazilian cities. This process transformed the structure of the social and spatial division of urban centers, as well as directly affecting the ways in which the population worked, generating an agglomeration of the population that ended up multiplying the points of concentration and production of the articulated and hierarchical urban network.

Brazil can be considered a country with late capitalism, due to its industrial backwardness as a colony, where it first served the capitalist interests of Portugal, England and the United States. This backwardness hindered Brazil in the race towards capitalism, as it created a lack of choice as to which equipment would be used in a given industrial area, mainly due to the lack of technology that could be imported, and when it was available, much of it was already second-hand.

A decisive point for Brazil's industrialization was the migration from slave to free labour, which was mainly carried out by European immigrants who arrived at the beginning of the 20th century, and by former slaves, as it took the country to a new stage of capitalism.

An important milestone in this new phase was the First World War, when European countries were forced to buy much-needed Brazilian products.

With this process of capitalist expansion at the beginning of the 20th century, and intense urbanization, especially in the major Brazilian capitals, the bourgeois and proletarian industrial classes emerged. These new classes united and began to question the sovereignty of the rural oligarchies. In addition, and with the support of the dictatorial power of President Getúlio Vargas, Brazilian industrialization was supported by an authoritarian regime and also by public capital.

Between the 1940s and 1950s, migration from the countryside to the city intensified, and the old agro-export society was transformed into a new urban-industrial society. This new form of spatial organization led to the shift of the hegemonic center of power to the Southeast. This region concentrated the great core of Brazilian industrialization and began to stimulate the modernization of the countryside.

The arrival of the military dictatorship was characterized as the era of the Brazilian economic miracle, in which countless changes took place in the national economy. The process of conservative modernization took place at the cost of a large foreign debt and a huge influx of foreign capitalism and technology. Despite this, workers suffered the most during this period, as the military regime prevented any kind of representation of the working class, due to its conservative reforms.

It was at this time that the incentive to export arose in order to compensate for the external and internal indebtedness that had been generated by investments in hydroelectric plants, urban infrastructure, among others.

2.2 Affordable housing in Brazil and its public policies

As we have already seen, the process of urbanization has triggered a fierce search for housing, thus generating housing that is segregated according to capitalist concepts. Based on this logic, and with the development of his skills, man began to develop his housing more and more. Despite this, shelter has not lost its main purpose, which is to protect the individual from the elements and from intruders.

In spite of this, according to Rapoport (1984 *apud* LARCHER, 2005), housing does not only have the function of providing shelter. It can be seen from the countless forms of construction that are located in the same place that it shows a human characteristic, where it is necessary to convey meanings of the differentiation of the inhabitants in relation to their neighbors and people outside their groups.

Nowadays, housing is a basic need, as well as being a project for any individual to acquire, and this is mainly due to capitalist logic (JUNQUEIRA and VITA, 2002). This factor justifies the large number of real estate investments being made in Brazilian cities.

Santos (1999) states that housing is a basic human need and aspiration. Home ownership, along with food and clothing, is the main investment in building up a patrimony, as well as being subjectively linked to economic success and a higher social position.

Despite these investments, a large part of the Brazilian population still lives in precarious situations or without housing.

The use of social housing began during the early days of industrialization, when workers who migrated from the countryside to the city in search of jobs in large industries received incentives from the government or large corporations to create housing areas, usually close to the industries themselves. Santa Maria is a landmark in this respect, where Vila Belga was built for the railway workers to live in with their families.

Between the 1930s and the 1960s, Brazil found itself under pressure to invest in social housing due to rising urbanization rates. State intervention in the housing sector began with the creation of the "Retirement and Pension Institutes" (IAPs) in the 1930s. These institutes began to finance housing for their members, through "building portfolios", thus generating an increase in the production of affordable housing units (FARAH, 1988 *apud* LARCHER, 2005).

In the post-war period, from 1945 onwards, the housing crisis in Brazil worsened as a result of the changes the country's industrial and agricultural economy underwent. A migratory movement from the more backward regions to the large urban centers began, along with the vegetative growth of the urban population, generating a great demand for new housing.

With this large urban agglomeration, and a large part of the population living in slums, alternatives were sought to get the population out of these areas and into areas of more decent housing.

In light of these facts, numerous public policies that have been implemented since the early 2000s are justified. These measures create housing of popular interest, benefiting a large portion of the population that does not have access to housing due to its high cost (ABIKO, 1995).

Pilar (2009) explains the reasons for this process when he says that despite the increase in civil construction in recent decades, it is only aimed at those who have the capital for housing. People who don't have the capital end up second-hand, waiting for housing policies, which are time-consuming and bureaucratic. During this process, this excluded population, with no housing options, ends up occupying areas that are unsuitable for housing, in areas that have little or no use.

This is justified by the words of Rodrigues (1997), who says that it is necessary to live, regardless of the location. Housing is a basic human need, just like clothing and food. The characteristics of housing change during the history of the human species, but it is still necessary to live, because it is not possible to live without taking up space.

Based on the words of Abiko (1995) who says that housing is the place occupied before and after the working day, where leisure and service tasks are accommodated, it is understood that housing must be a place where the basic principles of safety, housing and health are met.

Housing is a consumer good with unique characteristics, commonly found with a useful life span of over 50 years. Because it is an expensive consumer good, the less privileged classes constitute the greatest immediate demand for housing in the country (Fundaçâo Joâo Pinheiro, 2001).

Abiko (1995) divides social housing into three types:

- Low-income housing: this term implies the need to stipulate the maximum income for families and individuals in this bracket;

- Low-cost housing: cheap housing, but not necessarily aimed at the low-income population;

- Affordable Housing: generic term involving all solutions designed to meet housing needs.

Currently, the federal government stipulates that families participating in programs to build housing of popular interest must have an income of no more than R$ 1,600.00 per month.

It is currently clear that social housing units are financed by the public authorities, but can be carried out by private companies. In addition, it is aimed at people earning up to three times the minimum wage, as a process of inclusive action, but also as prevention of risk situations, environmental or cultural prevention in a given region.

"Social interest" as housing terminology in Brazil was already used in the programs for lower income groups of the now defunct National Housing Bank (ABIKO, 1995).

The solution to housing of popular interest goes far beyond simply building it. It is mainly linked to the income of the social classes benefiting from it, as well as the difficulties in accessing conventional financing credits, plus the deficiency in the implementation of housing policies (BRANDÂO, 1984 *apud* LARCHER, 2005).

According to Abiko (1995), popular housing should not be understood merely as a product, but as a process, with a physical dimension, but also as the result of a complex production process with political, social, economic, legal, ecological and technological determinants.

The need for housing is seen as a factor in building human dignity. Good housing conditions contribute to the socio-economic development of a given region. Despite this, in Brazil, as we can see from the 2010 Brazilian Municipal Housing Deficit, the country has a shortage of almost seven million homes, approximately 12.1% of total housing.

Housing deficit means the need for a total replacement of housing units and to meet the demand that is not solvable under the given market conditions (Fundaçâo Joâo Pinheiro, 1995).

Numerous factors influence the longevity of housing, making it a key aspect in defining the sustainability requirements of a social housing project.

One of these factors is incompatibility with the individual needs of each person. One of the most common of these factors is small-scale housing, which leads to expansion on the part of the residents themselves. Taking into account the expansion of the family, the lack of income for the expansion of the area of the residence generates a very large density in the bedrooms and ends up damaging the stay in the residential unit.

At a national level, the percentage of expansions in social housing is notable. This is mainly due to the expansion of families and the stagnation in the size of the homes built.

Due to the differentiation of each housing complex, some allow for greater availability for expansion, as is the case with plots with detached houses, either provided for in the project or as a solution adopted by the resident (LAUCHER, 2005).

On the subject of expansion, Laucher (2005) shows the priorities for expansion according to residents:

The expansion of the kitchen as a priority demonstrates the Brazilian culture, inherited from the Italian culture, where family gatherings revolve around cooking and the preparation process.

This expansion by the residents can be taken into account in the original design of the dwelling. To do so, sufficient private open spaces should be provided and the occupation of these spaces should be suggested in such a way that their use is preserved and that the internal spaces respond to the needs of the residents.

However, Reis (2002) states that the possibility of adapting projects to variations over time has not been adopted for detached houses, thus creating projects that do not respond to the needs of their users, as they do not have original layouts for this purpose.

3 SOCIO-SPATIAL FORMATION OF SANTA MARIA

3.1 The historical process and socio-spatial formation of the municipality of Santa Maria

In order to understand the historical process of the formation of the municipality of Santa Maria, it is necessary to go back to 1777 when the Portuguese crown in partnership with the Spanish crown signed an agreement called the Preliminary Treaty of Reciprocal Retributions, whose main objective was to demarcate the dominions of the two countries in southern Brazil.

In 1787, the team that would be responsible for demarcating the boundary between Spain and Portugal arrived at the site where the municipality of Santa Maria is located today and set up camp. The entourage stayed there for a long time until the markings were finished. Soon afterwards, orders were given for the construction of barracks for the troops, offices for the technical commission, ranches for the officers and the indispensable chapel in keeping with the religious demands of the time.

The expedition remained in the town until mid-1801, when the caravan left for Porto Alegre, leaving the village of Santa Maria behind. From the camp, the municipality expanded along the road. Such was its importance that it gave its name to the town's main street.

Santa Maria continued to expand as a village until it was named a town in 1858.

The city's expansion process was mainly due to the era of the railroads, where it was a hub because it was a transit area for trains that crossed the state and arrived from other countries with important cargo. The city developed around Avenida Rio Branco, supported by the economy that the railroad brought to the region.

The municipality did not have any foreign colonization process in its formation, but it did have ethnic miscegenation.

After the railroad, the establishment of the barracks and the Federal University of Santa Maria was what drove the expansion and formation of the municipality, making it a national reference in defense matters, with the second largest military contingent in the country, and home to numerous higher education institutions.

3.2 Santa Maria's socio-economic and natural characteristics

As mentioned earlier, the basis of the town's economy revolves around commerce and the civil service, which justifies the town's GDP of approximately R$ 3,855,271.743 thousand (IBGE, 2008).

Also according to the IBGE (2010), the municipality of Santa Maria/RS has a total area of 1788.1 Km^2 and has around 261,031 inhabitants, the fifth largest municipality in Rio Grande do Sul, behind only Porto Alegre (1450,555), Caxias do Sul (435,564), Pelotas (328,275) and Canoas (323,827). 95% of the population live in urban areas.

This population is divided into the 10 districts that make up the municipality, which are: Boca do Monte, Sâo Valetim, Santo Antâo, Arroio Grande, Palma, Arroio do Só, Passo do Verde, Santa Flora, Pains and the main district. It is further subdivided into 41 neighborhoods.

The city is located in a geological and geomorphological transition zone. The urban area developed on gentle hills rests on sandy-clay soils from the upper Triassic. To the north of the city begins the edge of Brazil's Southern Plateau, made up of basaltic soils with some intercalations of baked sandstone, the whole resting on aeolian sandstones of the Botucatu formation. From the edge of the plateau to the north, the regional name is "serra", although the landscape at the top is also made up of gentle hills. The district of Sede lies at an altitude of between 120 and 159 meters. The rugged topography of the rim, with various residual reliefs and valleys, justifies the local name of "serra" given to the Plateau region.

The municipality's climate is humid subtropical, with summers with very high temperatures of around 35°C and very cold winters, easily reaching temperatures below 0°C and with a lot of frosts. The average annual temperature in Santa Maria is 18.8°C. The warmest month is January and the coldest is July. Rainfall is well distributed throughout the year, and the annual average is 1,617 millimeters (mm).

Santa Maria, due to its central geographical position and its location in the southern half of the state, has been historically strategic since the time of the empire in terms of conflicts with the "silver countries". For this reason, for several decades the investments concentrated there were related to national security.

This led to the formation of a structure and an economic vocation for the municipality focused on the provision of services, which was later accentuated with the establishment of state and federal public services and the development of commerce.

The economic foundations of the municipality can be seen in the jobs on offer. The available data reveals the high importance of the tertiary sector, with trade, public services, including those of the Federal University of Santa Maria, and the military standing out.

Certainly, the large mass and flow of money in the city of Santa Maria depend fundamentally on public services. As has already been pointed out, Santa Maria stands out in the region, the state and the country as a city with the following functions related to the provision of services: commercial, educational, medical, hospital, road and military police.

These tertiary urban functions absorb more than 80% of the city's active population, with the commercial and educational sectors standing out. Still in the functional aspect of the city, the primary sector (agriculture and livestock) is in second place and the secondary sector is in third place, which in general are small and medium-sized industries, mainly focused on the processing of agricultural products, metallurgy, furniture, footwear, dairies, etc.

The city stands out for being the second city in Rio Grande do Sul in terms of the number of wealthy people, and the second city in the state with the highest number of people in classes A and B (28 in the country). According to a survey by the Getúlio Vargas Foundation.

4 ANALYSIS AND INTERPRETATION OF RESULTS

This chapter aims to analyze and interpret the results obtained in the research.

4.1 Affordable housing in Santa Maria/Rs

As has already been mentioned, the area under study is an important center of real estate pressure, where numerous developers invest in high-end condominiums, where a small part of the population has access to these homes, thus generating great housing inequality nationwide. In order to remedy these social inequalities and the Brazilian housing deficit, numerous housing policies have been implemented in recent years. In this way, various programs have been disseminated throughout Brazil, benefiting a huge number of families in all regions of the country. The main objective of these programs, which are offered by the federal government, is to provide dignified access to financing and, consequently, to purchase a home. The municipality of Santa Maria is not far from this reality, and has been receiving incentives from the federal government to set up these low-cost housing units.

Firstly, the municipality was awarded the Growth Acceleration Program (PAC-Habitacional), which consists of a planning, management and execution model for public investment, which combines public and private infrastructure projects as well as institutional measures to increase the pace of economic growth (BRASIL, 2007).

The PAC, set up in 2007 by the federal government, was divided into two stages, called PAC 1 and PAC 2, and generated countless projects at national and municipal level. In Santa Maria, it developed the construction of three housing estates, before being instigated in 2014. The three housing estates delivered to the municipality by PAC-Habitaçâo are: Residencial Cripriano da Rocha; Residencial Loteamento da Brenner and Residencial Lorenzi. Together, they total approximately 1,000 homes delivered to low-income families in Santa Maria.

With the extinction of PAC-Habitaçâo, the federal government invested in the Minha Casa Minha Vida Program, in partnership with Caixa Econômica Federal, as well as state governments and municipal governments.

The MCMV was implemented in Santa Maria in 2011, with the construction of the program's first housing complex in the municipality. There are currently three homes already delivered in the territorial unit, and a fourth is in the process of being finalized, and is expected to be delivered to the recipients in January 2016, according to the Santa Maria City Hall.

As has already been mentioned, there are three residential units delivered in Santa Maria, and one in the process of being delivered. Currently, the municipality has the Residencial Videiras, delivered in October 2011, with 420 apartments; the Residencial Zilda Arns, delivered in June 2012, with 500 apartments; the Residencial Dom Ivo Lorscheiter, completed in December 2014, with 580 apartments, and the Residencial Leonel de Moura Brizola (January 2016) with 362 apartments.

4.2 Selection and selection process for families

The aforementioned housing estates were set up through the popular housing program, financed by the Federal Government in conjunction with the Santa Maria/RS City Hall. The target audience of this program is families with an income of less than R$ 1,600.00. The program has a detailed selection process, and the state or municipality is responsible for nominating the families that will be eligible.

An important factor that the residents were asked about was how they found out about the program. The vast majority said it was through meetings, where Santa Maria councillors went to low-income communities to present the projects. Others said that they found out about the program through television and other media. They also pointed out that there is a need for more publicity about the means of registration and selection for the program.

The process for selecting families for the program is very strict and follows national parameters. In addition, each municipality has the freedom and authority to create its own criteria for selecting beneficiaries. At the national level, the prerequisites for a family to register for the housing complexes are divided by priority, where they are prioritized as follows: firstly, families living in risky or unhealthy areas; secondly, families with women responsible for the family unit; finally, families that include people with disabilities. At the municipal level, applicants must meet the following criteria: families with children under the age of 18; families with 3 or more children under the age of 18; the head of the family must be a beneficiary of the Bolsa Familia program.

In the Territorial Unit under study, the Municipal Department for Social Development is responsible for registering those interested. According to the secretary in charge of the department, the registration process is quite simple. Interested families first need to go to the Town Hall with the following documents: CPF and ID card; work permit; updated SIBEC report and/or current Bolsa Familia payment statement; proof of residence (with zip code); proof of income; children's birth certificates; electoral certificate and/or school certificate and/or registration with utilities; families living in areas at risk must present a Civil Defense certificate of occurrence and a copy of the document; families with people with disabilities must present a medical certificate proving the type, degree or level of the alleged disability and the International Classification of Diseases (ICD).

In addition, according to the Secretary, the registered families go through a rigorous committee to evaluate the information provided. This committee is made up of social workers employed by the town hall. In addition to the bureaucratic part, the selection process involves an interview and a visit from the social workers. These officials go to the current home of the applicant family and get to know the reality of the family, as well as understanding the needs of the applicants.

As for the amounts paid by the recipients, there was a difference between the first and most recent housing developments. A priori, the amount paid to the recipients was equivalent to 10% of the income reported by the applicants. More recently, the federal government revised these amounts and decided that they should be equivalent to 5% of the reported income. In order to ensure that all those who benefited from the program paid the same percentage, those who were already in their homes received discounts on the amount already paid. Currently, all residents pay the 5% stipulated, and these installments are between R$23.00 and R$74.00 per month, for 10 years. During this period, the houses cannot be rented, donated, given away or sold. According to the town hall, the total cost of the homes is around R$64,000.00.

As we can see, the selection process is very careful, but interesting, because it puts the bodies responsible in the reality of the beneficiaries, where there is a humanization of the problems, so that the applicant families are not just treated as numbers. According to the town hall agents, it is clear that these families see the possibility of housing as a way of acquiring dignity for their family, seeing this process of contemplation as a kind of fresh start.

However, during the field interviews, when residents were asked how they rated the process of registering and selecting beneficiaries, there were numerous complaints. Many residents claimed that they had applied several times before being selected. In addition, some also pointed out that: "I believe that in some cases it is necessary to have an acquaintance inside (the town hall), the famous QI (Quem Indica)". (Resident no. 1).

It can be seen, then, that despite the many positive points of the program, it still generates a certain amount of dissatisfaction and distrust among some of those selected, and this form of selection for affordable housing needs to be rethought.

4.3 Location of housing estates in Santa Maria

Due to its economy, Santa Maria has an extremely densely populated urban center, where the city center is the scene of large vertical constructions, both for housing and commercial purposes. Due to this densification, neighborhoods further away from the central region are developing, as is the case of Bairro Camobi and Bairro Tancredo Neves. According to the Santa Maria Municipal Department of Housing and Land Regularization, in the eastern part of the city alone there are 45 condominiums in the process of being built or cleared for construction. With the growth of high-end gated communities in the municipality, there are few places to build affordable housing.

The Minha Casa Minha Vida housing complexes in Santa Maria, with the exception of Residencial Videiras which is close to the city center, are basically located in the expanding region of the territorial unit under study, close to Bairro Sâo José, on the banks of the RSC-287. With this map, we can see the exact location of the residences in the urban perimeter of Santa Maria (MAP 2).

Map 2: Location of the Residences under study.

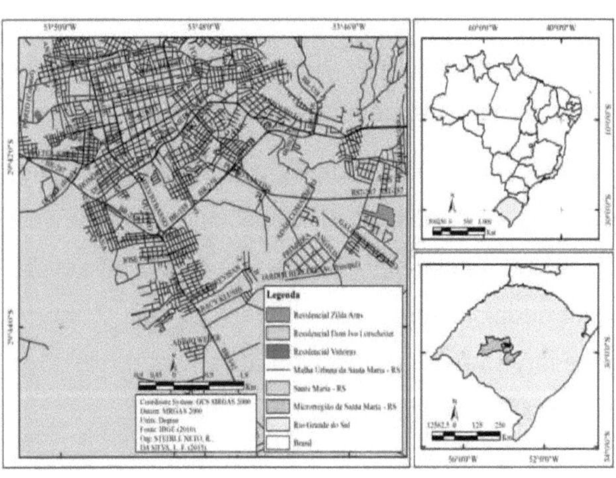

Source: IBGE, 2010.

4.3.1 Residencial Videiras

Residencial Videiras stands out from the rest, both because of its location and its structural form, as it is the only housing complex built in the form of a closed condominium, dividing the 420 apartments into 20 vertical blocks.

24

Located on the west side of the city, close to the center, more precisely in the Passo d'Areia district, on Rua Venâncio Aires, one of the municipality's main streets (Map 3), the residential development is close to a range of services and accessibility, which sets it apart from the other housing developments under analysis.

The neighborhood has an army barracks on its territory, where the Santa Maria Military College is located, as well as a Brigada Militar barracks.

It also has numerous schools and health facilities, such as the Santa Maria Garrison Hospital (HGuSM).

With a population of 6995 and an area of 2.6781 km^2 , it has a population density of 2611 inhabitants per km2. It is the 14th most populous neighborhood in the municipality, with a predominance of female residents (IBGE, 2010).

The neighborhood has been developing in recent years, especially after the completion of the Perimetral Avenue, which connects the Midwest region with the northern part of the city. The work, which began in 2008 and ended in 2013, revitalized the banks of the Cadena stream, creating paved spaces with cycle and walking areas. With the completion of the avenue, numerous developments have been built in the area. The most notable is a store belonging to the global Walmart chain, the Big Hypermarket, which opened at the end of 2012.

According to the town hall, the site was chosen for the residential development due to its privileged location and the services available close by, which provide great convenience and satisfaction for the residents.

According to the municipal secretary for social development, a study was carried out for the implementation of Residencial Videiras, and as it is a vertical construction project, it didn't require such a large area for construction. The city council then decided to build the housing complex.

Map 3: Location of Residencial Videiras.

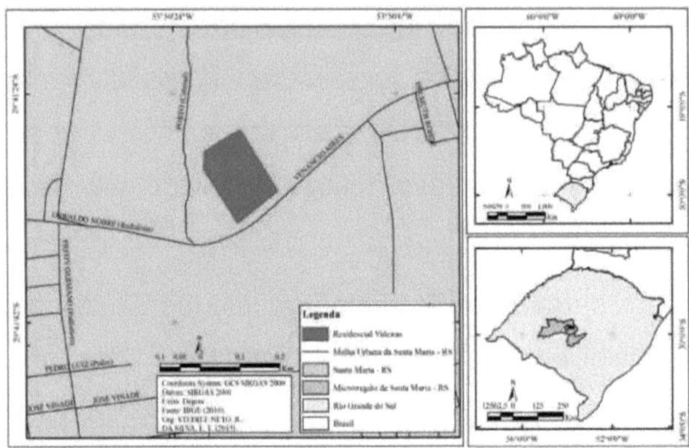

Source: IBGE, 2010.

4.3.2 Zilda Arns and Dom Ivo Lorscheiter Residences

Delivered in July 2012 and December 2014, respectively, Residencial Zilda Arns (Map 4) and Residencial Dom Ivo Lorscheiter (Map 5) were the second and third MCMV housing developments delivered in the municipality of Santa Maria. They have 500 and 578 residential units, respectively, divided into semi-detached houses.

Unlike Residencial Videiras, the residences were built in an open condominium, where anyone can access the residence and the houses, as they were delivered without protective fences.

Located in the Diàcono Joâo Luiz Pozzobon neighborhood, on the banks of the RSC-287, the main link between the central region of the state and the capital. This is a recent neighborhood, founded in 2006 in honor of Joâo Luiz Pozzobon. Its area is the sum of the entire area of the old cerrito, which until now was part of Bairro Sâo José.

As it is an area of urban expansion, there are many areas of countryside where livestock farming takes place, and there are also areas of agriculture.

In addition, the micro-basin of the Passo das

Troops.

Map 4: Location of Residencial Zilda Arns

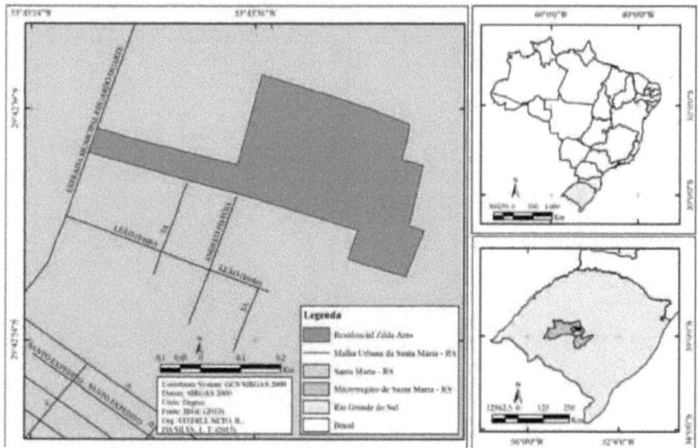

Source: IBGE, 2010.

With a population of 3152 inhabitants, spread over an area of 7.0884 km^2 , it has a population density of approximately 444 inhabitants per km2 (IBGE, 2010).

According to the town hall, the site was chosen mainly because of the space available in the area, as it was a housing complex with semi-detached houses, which required a larger area. However, it can be seen that the choice of this area did not completely meet the needs of the residents, as it is a remote neighborhood and difficult to access in some cases. Another factor that displeases the area is the issue of security. The residential area borders Vila Maringà, where most of the municipality's crime is concentrated.

Map 5: Location of Residencial Dom Ivo Lorscheiter

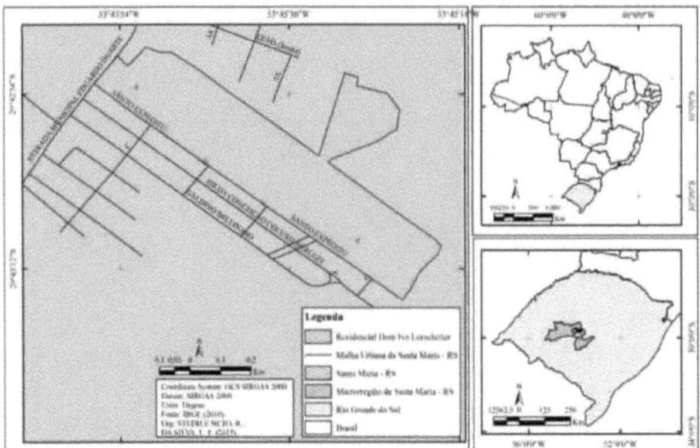

Source: IBGE, 2010.

4.4 Getting to know the socio-economic reality of the residences

Once we know the location of each residential unit, and the reasons why the city council chose a particular area for development, we can get a better idea of each housing complex. Once this part has been captured, we need to analyze the socio-economic reality in which the residential units are located, looking at their residents. To this end, questionnaires were administered to a number of residents in order to understand the reality in which they live.

4.4.1 Residencial Videiras

The Videiras residential complex, as already mentioned, was handed over to the recipients in 2011 as a closed condominium. The housing complex has 420 residential units and, according to data from Santa Maria City Hall, is home to approximately 1,000 people.

During the two days of fieldwork, 21 questionnaires were carried out with residents, and it can be seen that the average family income of residents is R$1,000.00, and that on average 3.3 people live in each home (TABLE 1). In addition, the level of schooling is low among those responsible for the family, but increases among young people. It was clear that one of the motivations for enrolling in the program was to provide decent places for their children to live and study.

In addition to the economic reality, the structure of the families living there can also be seen. According to the interviews, most families have only one responsible member, and many have grandparents. This data is worrying, as it demonstrates the breakdown of the family, bringing countless benefits for young people.

As already mentioned, a positive point mentioned by many residents was education. Although the level of education among those responsible is low, young people are seeking to improve themselves through study. According to the residents, taking part in this program favors their children's education. Another point made was the proximity to schools. In their former homes, the children had to travel in the rain, often through bumpy and muddy streets, and often on the most severe stormy days, it was not possible for them to get to school. What's more, there are countless accounts of the effort these families make to educate their youngsters. This is justified by the words of one of the residents when he says that:

"My family has always been very poor. I had to work with my parents from the age of eight. When I got married, I always thought about giving my children a different life. Thank God we were lucky enough to win the houses. Although I've always had a very simple life, my children's studies have always been a priority. I work from Monday to Monday so that they never miss anything, and I know that they value this very much and repay me with their achievements at school. My children are excellent students and I'm very proud of my role as a mother." (Resident no. 6)

Table 1: Identification of the residents interviewed:

Resident identification	Family members	Age	Education	Family income
Resident No. 1	Mother; Father; 2 children	Mother: 43 years old Father: 47 years old Children: 13 and 17 years old	Mother: Primary school; Father: 5th grade[a] ; Children: studying	R$ 950,00
Resident no. 2	Mother; 3 children	Mother: 37 years old Children: 5, 9 and 14 years old	Mother: Secondary school incomplete Children: studying	R$ 788,00
Resident no. 3	Father; Mother; 1 child	Father: 41 years old; Mother: 43 years old Son: 18 years old	Complete secondary education	R$ 1.200,00

Resident no. 4	Husband and wife	Husband: 73 years old; Wife: 69 years old.	5th grade elementary school	R$ 788,00
Resident no. 5	Father; 1 son	Father: 30 years old; Son: 8 years old	Father: completed secondary school; Son: studying	R$ 1.000,00
Resident no. 6	Father; Mother; 2 children	Father: 45 years old; Mother: 45 years old; Children: 18 and 15 years old	Father: Elementary School Mother: Middle School; Children: Middle School	R$ 830,00
Resident no. 7	Mother; Grandmother; 2 children	Grandmother: 77 years old; Mother: 28 years old; Children: 3 and 5 years old	Grandmother: 4th grade. Elementary Mother: Med.	R$ 800,00
Resident no. 8	Grandmother; 3 grandchildren	Grandmother: 65 years old; Grandchildren: 27, 20 and 15 years old	Grandmother: Elementary School Grandchildren: Med.	R$ 788,00
Resident no. 9	Mother; 4 children	Mother: 37 years old; Children: 17, 15, 12 and 7 years old	Mother: Elementary school Children: studying	R$ 600,00
Resident no. 10	Mother; 1 son	Mother: 45 years old; Son: 24 years old	Mother: Secondary Education; Son: Higher Education	R$ 788,00
Resident no. 11	Mother; Father; 2 children	Mother: 34 years old; Father: 37 years old; Children: 18 and 13 years old.	Mother: Med. teaching; Father: Med. teaching Children: studying	R$ 1.500,00
Resident no. 12	Grandmother; 1 grandchild; 1 great-grandchild	Grandmother: 64 years old; Grandson: 21 years; Great-grandson: 08 months	Grandmother: Illiterate; Grandson: Med. Ed.	R$ 788,00
Resident no. 13	Husband and wife	Husband: 77 years old; Wife: 69 years old.	Elementary School	R$ 788,00
Resident no. 14	Father; Mother; 1 child	Father: 33 years old; Mother: 33 years old; Son: 4 years old	Father: Medical School; Mother: Medical School.	R$ 1.000,00
Resident no. 15	Grandmother; 2 grandchildren	Grandmother: 62 years old; Grandchildren: 21 and 19 years old	Grandmother: Elementary School Grandchildren: Ens.	R$ 1.300,00

			Med.	
Resident no. 16	Mother; Father; 3 children	Mother: 37 years old; Father: 39 years old; Children: 15, 13 and 10 years old	Mother and Father: Middle School; Children: studying	R$ 1.600,00
Resident no. 17	Mother; 1 son	Mother: 40 years old; Son: 21 years old	Mother: 5ª series; Son: ens. Med.	R$ 560,00
Resident no. 18	Father; Son; Daughter-in-law; Grandson	Father: 39 years old; Son: 18 years old; Daughter-in-law: 17 years old; Grandson: 1 year old	Father: 7th grade; Son: 2nd grade. Med. Daughter-in-law: Primary Education	R$ 950,00
Resident no. 19	Mother; 1 son	Mother: 45 years old; Son: 19 years old	Mae: Medical Education Son: Med.	R$ 1.600,00
Resident no. 20	Father; Mother; 3 Children	Father: 57 years old; Mother: 50 years old; Children: 21, 18 and 16 years old	High School	R$ 1.500,00
Resident no. 21	Mother; 2 children	Mother: 30 years old; Children: 15 and 10 years old	Mother: High School; Children: studying	R$ 900,00

Source: Fieldwork, 2015.

Org.: STEDILE NETO, R. 2015.

There have also been countless improvements in the labor market. Many people say that before affordable housing, it was very difficult to get a job due to the lack of legalized housing. Some residents reported that in their old homes, because they were areas that didn't have a zip code, places that weren't legalized by the government, they were unemployed for years, because most employers required their employees to have a fixed address, making it difficult for the low-income population. This can be seen in the words of one of the interviewees, who said: "Before I moved into my own home, I was unemployed for seven years. I lived in Brenner, an area that wasn't legalized and I didn't have proof of residence. After I got the house, I got a job and I'm very happy" (Resident no. 19).

Based on this question, a table was drawn up with the professions and number of dependents per household of the residents interviewed (TABLE 2).

Table 2: Economic table of the interviewees (V).

Resident Identification	Responsible Member's Profession	Salary	No. of dependents
Resident No. 1	Market Box	R$ 950,00	3
Resident no. 2	Saleswoman	R$ 788,00	3
Resident no. 3	Bus collector	R$ 1,200.00	2
Resident no. 4	Retired	R$ 788,00	1
Resident no. 5	Civil servant	R$ 1,000.00	1
Resident no. 6	Stonemason	R$ 830,00	3
Resident no. 7	Saleswoman	R$ 800,00	3
Resident no. 8	Retired	R$ 788,00	3
Resident no. 9	Domestic	R$ 600,00	4
Resident no. 10	Saleswoman	R$ 788,00	1
Resident no. 11	Manicurist/stoner	R$ 1,500.00	3
Resident no. 12	Retired	R$ 788,00	2
Resident no. 13	Retired	R$ 788,00	1
Resident no. 14	Stonemason	R$ 1,000.00	2
Resident no. 15	Bus collector	R$ 1,300.00	2
Resident no. 16	Self-employed	R$ 1,600.00	4
Resident no. 17	Cleaning lady	R$ 560,00	1
Resident no. 18	Stonemason	R$ 950,00	3
Resident no. 19	Teacher	R$ 1,600.00	1
Resident no. 20	Civil servant	R$ 1,500.00	4
Resident no. 21	Hairdresser	R$ 900,00	2

Source: Fieldwork, 2015.

Org.: STEDILE NETO, R. 2015.

As we can see from the table above, 96% of households have the income of just one family member. The number of young people working to help with the family income was also absent. All those interviewed said that their children's education was a priority and that it would be a distraction if they worked.

Another factor asked about was the value of the installments, which average R$50 a month. According to the residents, the amount is good and fits perfectly into their budget. They also pointed out that if it weren't for these programs they would never be able to own their own home. They treat housing stability as a very important factor because they "feel good, comfortable in a place they know is ours". (Resident no. 21).

4.4.2 Zilda Arns Residence

The second MCMV housing project to be built in Santa Maria, it was completed in June 2012 and handed over to its beneficiaries. Unlike Residencial Videiras, Zilda Arns was built as an open condominium with semi-detached houses. It has 500 residential units and is home to approximately 1,500 people, according to Santa Maria's Social Development Department.

During the fieldwork, 25 interviews were conducted during the morning shift. It can be seen that the average family income of the residents is R$835.20 and on average 2.7 people live in each home (TABLE 3). The educational level of the residents was also surveyed and, as in the previous housing estate, the level is low among the older residents and increases among the younger ones.

As we can see, the housing estate is not far behind the others, with a low family income and a low level of schooling among the older residents. We can also see that 100% of young people of school age are enrolled, according to the residents themselves.

Table 3: Identification of the residents interviewed in the Zilda Arns Residence.

Resident identification	Family members	Age	Education	Family income
Resident No. 1	Mother; Son; Granddaughter	Mother: 68; Son: 40; Granddaughter: 17	Mother: grade 6ª ; Son: high school Granddaughter: studying	R$ 788.00
Resident no. 2	Husband and wife	Husband: 54; Wife: 51	Medical Education Inc.	R$ 1.500,00
Resident no. 3	Father; Mother; Son	Father: 70; Mother: 65; Son: 21	Father and Mother: 5th grade; Son: 7th grade	R$ 950,00
Resident no. 4	Mother; 3 children	Mother: 37; Children: 13, 9 and 5	Mother: 7th grade; Children: studying	R$ 1.200,00
Resident no. 5	Father; Mother; 2 children	Father: 48; Mother: 44; Children: 18 and 13	Father, mother and eldest son: Med. ed. Youngest son: Ens. Fund.	R$ 788,00
Resident no. 6	Grandfather; Daughter; Grandson	Grandfather: 58 years old; Daughter: 25 years old; Grandson: 5 years	Grandfather: 4ª grade; Daughter: Elementary school.	R$ 680,00

		old		
Resident no. 7	Father; Mother; Son	Father: 39; Mother: 37; Son: 5	Father: Childcare Technician Mother: Med. Ed.	R$ 1.600,00
Resident no. 8	Husband and wife	Husband: 72; Wife: 75	3rd grade	R$ 540,00
Resident no. 9	Mother; 2 children	Mother: 27; Sons: 2 and 3.	Elementary school	R$ 500,00
Resident no. 10	Father; Mother; 3 children	Father: 42; Mother: 40; Children: 15, 10, 3	Father and mother: Elementary school Children: studying	R$ 830,00
Resident no. 11	Mother; 3 children	Mother: 32; Children: 10, 10 e 5	Mother: Teaching Med. Children: studying	R$ 950,00
Resident no. 12	Husband and wife	Husband: 50; Wife: 49	High School	R$ 1.150,00
Resident no. 13	Owner	47	Elementary School	R$ 788,00
Resident no. 14	Father; 1 son	Father: 34; Son: 8	Father: Elementary School	R$ 500,00
Resident no. 15	Grandmother; 2 grandchildren	Grandmother: 56; Grandchildren: 4 e 2	Primary Education	R$ 550,00
Resident no. 16	Mother; Father; 2 children	Mother: 37; Father: 33; Children: 10 and 6	Father and mother: teaching medicine Children: studying	R$ 640,00
Resident no. 17	Owner	59	Elementary School	R$ 788,00
Resident no. 18	Mother; 1 son	Mother: 33; Son: 2	Elementary School	R$ 600,00
Resident no. 19	Father; 2 children	Father: 45; Children: 19 e 15	Father: Med. Children: studying	R$ 900,00
Resident no. 20	Mother; Father; 3 children	Mother: 40; Father: 40; Children: 12, 10, 7	Mother and Father: Elementary School Children: Studying	R$ 500,00
Resident no. 21	Husband and wife	Husband: 30; Wife; 25	Elementary School	R$ 800,00
Resident no. 22	Husband and wife	Husband: 71; Wife: 70	5th grade elementary school	R$ 788,00
Resident no. 23	Grandmother; granddaughter	Grandmother: 61; Granddaughter: 17	Grandmother: 6th grade; Granddaughter: Med.	R$ 700,00
Resident no. 24	Husband; Wife	Husband: 36; Wife: 35	Elementary School	R$ 650,00

| Resident no. 25 | Owner | 53 | | High School | R$ 700,00 |

Source: Fieldwork, 2015.

Org.: STEDILE NETO, R. 2015.

The structure of families is quite striking, as it can be seen that around 20% of families have only one responsible member. It is clear that many family units are broken up by violence and crime. Proof of this can be found in the words of one of the residents:

"I raise my two grandchildren alone. My daughter-in-law left the children for me and my son to raise and disappeared into the world. When I heard about these houses I signed up, I saw here a good place to raise my children. I had a son die last year (2014), he was a drug user. He was always a great father, I fought for him, but that's life. I'm raising my grandchildren, and I know they'll be good people." (Resident 15).

With these words, the social issue in which the housing complex is inserted becomes clear. Many of these families were surrounded by violence and numerous social problems and they see these housing policies as a way out of this reality.

It also made it possible for us to draw up an economic profile of the residents of this housing estate (TABLE 4). Obtaining this data made it easier to understand the economic reality of the beneficiaries.

Table 4: Economic table of the interviewees (Z.A.).

Resident identification	Responsible Member's Profession	Salary	No. of dependents
Resident No. 1	Retired	R$ 788,00	2
Resident no. 2	Retired	R$ 1.500,00	1
Resident no. 3	Seller	R$ 950,00	2
Resident no. 4	Snack	R$ 1.200,00	3
Resident no. 5	Stonemason	R$ 788,00	3
Resident no. 6	Autonomous	R$ 680,00	2
Resident no. 7	Computer Technician	R$ 1.600,00	2
Resident no. 8	General services	R$ 540,00	1
Resident no. 9	Collector	R$ 500,00	2
Resident no. 10	Saleswoman	R$ 830,00	3
Resident no. 11	Hairdresser	R$ 950,00	3
Resident no. 12	Artisans	R$ 1.150,00	1
Resident no. 13	Domestic	R$ 788,00	-
Resident no. 14	General Services	R$ 500,00	1
Resident no. 15	Domestic	R$ 550,00	2
Resident no. 16	Hatter	R$ 640,00	3
Resident no. 17	Retired	R$ 788,00	-

Resident no. 18	Diarist	R$ 600,00	1
Resident no. 19	Stonemason	R$ 900,00	2
Resident no. 20	Unemployed	R$ 500,00	4
Resident no. 21	Electrician	R$ 800,00	1
Resident no. 22	Retired	R$ 788,00	1
Resident no. 23	Retired	R$ 700,00	1
Resident no. 24	Worker	R$ 650,00	1
Resident no. 25	Arts	R$ 700,00	-

Source: Fieldwork, 2015.

Based on this data, we can see that the employment relationship of the residential population is that of service providers or INSS pensioners. We can also see that most families are characterized by only one person being responsible for the family income. There were also no reports of young people helping to generate income.

When asked how the contemplation had changed the family's financial life, many reported that by no longer having to pay rent they were able to invest more in comfort, thus increasing the quality of life for all the members. It is also worth mentioning that all the residents of the housing complexes are or were beneficiaries of the Bolsa Familia program.

4.4.3 Dom Ivo Lorscheiter Residence

The third housing complex delivered to Santa Maria was the Dom Ivo Lorscheiter Residences, which were handed over to the recipients in 2014. Like Zilda Arns, this housing development was built as an open condominium and has 578 semi-detached houses. According to the City Hall, it houses around 1,800 people.

The fieldwork to obtain the data took place over two days, and 28 questionnaires were carried out with the local residents. From the results obtained during the fieldwork, it was possible to conclude that the average family income is R$919.50 and that an average of 2.7 people live in each home (TABLE 5). Not far from the reality of the other residences, Dom Ivo Lorscheiter has a low level of schooling among older people, and a significant increase among younger people.

Table 5: Identification of Interviewees at Dom Ivo Lorscheiter

Resident identification	Family members	Age	Education	Family income
Resident No. 1	Father; Stepmother; 2 children	Father: 32 years old; Stepmother: 27 years old; Children: 5 and 3 years old	Elementary School	R$ 980,00
Resident no. 2	Father; Mother; 1 child	Father: 35; Mother: 35; Son: 13	Father and Mother: Teaching Medicine Son: studying	R$ 1.600,00
Resident no. 3	Mother; 2 children	Mother: 45; Children: 18 e 15	Mother: Med. Children: Ens. Med. and studying	R$ 950,00
Resident no. 4	Husband and wife	Husband: 68; Wife: 63	Elementary School	R$ 788,00
Resident no. 5	Mother; 3 children	Mother: 37; Children: 5, 3 e 1	High School	R$ 850,00
Resident no. 6	Father; Mother; 2 children	Father: 29; Mother: 28; Children: 2	High School	R$ 900,00
Resident no. 7	Father; 1 son	Father: 40; Son 20	High School	R$ 1.600,00
Resident no. 8	Grandmother; Granddaughter	Grandmother: 70; Granddaughter: 15	Grandmother: 5ª série; Granddaughter: studying	R$ 788,00
Resident no. 9	Mother: 2 children	Mother: 22; Children: 3 and 1	High School	R$ 600,00
Resident no. 10	Father; Mother; 1 child	Father: 27; Mother: 28; Son: 5	Elementary School	R$ 800,00
Resident no. 11	Husband and wife	Husband: 48; Wife: 45	High School	R$ 1.500,00
Resident no. 12	Husband; Wife	77	5th grade	R$ 788,00
Resident no. 13	Mother; 1 son	Mother: 40; Son: 12	High school; studying	R$ 650,00
Resident no. 14	Father; Mother; 2 children	Father: 57; Mother: 60; Sons: 21 and 15	Father and Mother: Primary Education Children: Higher Inc. and Fund.	R$ 1.200,00
Resident no. 15	Father; 2 children	Father: 37; Sons: 10 and 8	Father: Tech. Children: studying	R$ 1.000,00
Resident no. 16	Father; Mother; Son	Father: 40; Mother: 40; Son: 10	Med. teaching; studying	R$ 788,00
Resident no. 17	Mother; 3 children	Mother: 35;	Elementary School	R$ 900,00

		Children: 7, 5 e 3		
Resident no. 18	Father; 1 Son	Father: 27; Son: 5	Elementary School	R$ 700,00
Resident no. 19	Grandfather; grandson	1 Grandmother: 58; Grandson: 16	Grandmother: 6th grade; Neto: studying	R$ 788,00
Resident no. 20	Mother; 1 son; nephew	1 Mother: 35; Son: 10; Nephew: 16	Mother: Elementary School Son and nephew: studying	R$ 800,00
Resident no. 21	Mother; 2 children	Mother: 49; Children: 19 e 15	Mother: Teaching Med. Children: studying	R$ 950,00
Resident no. 22	Father; Mother; 2 children	Father: 50; Mother: 47; Sons: 17 and 15	Father and mother: Ens High School; Children: studying	R$ 1.000,00
Resident no. 23	Husband; Wife	65	7ª series	R$ 600,00
Resident no. 24	Father; 2 children	Father: 23; Sons 3 and 1	Elementary School	R$ 750,00
Resident no. 25	Mother; 1 son	Mother: 51; Son; 24	Mother: Primary Education Son: Med.	R$ 700,00
Resident no. 26	Grandmother; grandchildren	3 Grandmother: 60; Grandchildren: 15, 12 e 7	Grandmother: 5th grade; Grandchildren: studying	R$ 788,00
Resident no. 27	Grandmother; grandchild	1 Grandmother: 63; Grandson: 20	Grandmother: Elementary School Neto: Technical course	R$ 788,00
Resident no. 28	Owner	56	High School	R$ 1.200,00

Source: Fieldwork, 2015.

Org.: STEDILE NETO, R. 2015.

We can see that Residencial Dom Ivo Lorscheiter does not differ from the others in terms of education; the level of education among the elderly is low, and the number of people with higher education is non-existent. However, it can be seen that the young people are part of a school reality and are looking for a better future for their families.

One can see the satisfaction and happiness among those interviewed at having their own place to live. According to the residents, the place is very quiet and pleasant to live in, and the families' income has increased as they no longer have to pay rent. One of the residents says

"Before I moved here, half the family's income went on rent, but nowadays we can buy appliances and other things to complement our comfort in the house. There's no doubt that the place needs a lot of improvements, but for those who had nothing, having their own place is fantastic." (Resident no. 17).

The interviews were used to create an economic profile of the residents, where it was possible to survey the salaries and occupations of each head of household (TABLE 6).

Table 6: Economic table of the interviewees (D.I.L.)

Resident identification	Responsible Member's Profession	Salary	No. of dependents
Resident No. 1	Stonemason	R$ 980,00	3
Resident no. 2	Bus driver	R$ 1.600,00	2
Resident no. 3	Manicure	R$ 950,00	2
Resident no. 4	Retired	R$ 788,00	1
Resident no. 5	Cleaning lady	R$ 850,00	3
Resident no. 6	Stonemason	R$ 900,00	3
Resident no. 7	Taxi driver	R$ 1.600,00	1
Resident no. 8	Retired	R$ 788,00	1
Resident no. 9	Autonomous	R$ 600,00	2
Resident no. 10	Self-employed	R$ 800,00	2
Resident no. 11	Store manager	R$ 1.500,00	1
Resident no. 12	Retired	R$ 788,00	1
Resident no. 13	Diarist	R$ 650,00	1
Resident no. 14	Saleswoman	R$ 1.200,00	3
Resident no. 15	Stonemason	R$ 1.000,00	2
Resident no. 16	Mechanic	R$ 788,00	2
Resident no. 17	Hairdresser	R$ 900,00	3
Resident no. 18	Stonemason	R$ 700,00	1
Resident no. 19	Retired	R$ 788,00	1
Resident no. 20	Cleaner	R$ 800,00	2
Resident 21	Epilator	R$ 950,00	2
Resident no. 22	Mechanic	R$ 1.000,00	3
Resident no. 23	Self-employed	R$ 600,00	1
Resident no. 24	Stonemason	R$ 750,00	2
Resident no. 25	Diarist	R$ 700,00	1
Resident no. 26	Retired	R$ 788,00	3
Resident no. 27	Retired	R$ 788,00	1
Resident no. 28	Snack	R$ 1.200,00	-

Source: Fieldwork, 2015.

Org.: STEDILE NETO, R. 2015.

Based on the data obtained in this economic table, it can be seen that the average income is low, and that in the vast majority only one member of the family is responsible for the entire income. There is also a lack of young people working to contribute to the family income. It can also be seen that the number of INSS pensioners is also high, and that a large proportion of families are run by grandparents.

According to the residents, although they have only been living there for a short time (about a year), the changes in their lives are noticeable. Approximately 40% of those interviewed reported that they had left unemployment after moving, as they were able to declare a fixed address to their employers.

Reports like this show the importance of policies like these, which are increasingly improving the quality of life of the Brazilian population.

4.5 Physical infrastructure of the residences

After understanding the social structure of the housing estates, we also need to know the physical infrastructure of each housing estate. To do this, we took into account the area of the house, the social areas present in the condominium and the health, education, transportation and security issues. The data was obtained during fieldwork and from consultations with specialized agencies such as the Santa Maria City Hall and Caixa Econômica Federal.

4.5.1 House area

It's well known that modern housing is getting smaller and smaller due to strong real estate pressure and the maximization of profits for construction companies. Large high-end residential projects are characterized by houses built on the basis of 40 m^2 , thus increasing the number of residential units for sale or rent.

Not far from this reality, the housing complexes of the Minha Casa Minha Vida program are also built on a small base. This built-up area is defined by the parameters of the program, which are defined by Caixa Econômica Federal, and differ in terms of the type of housing.

For the construction of apartments, which is exclusively the case with Residencial Videiras, the house must contain: two bedrooms, a living room, kitchen, bathroom and laundry area, making a total of six rooms. They must be divided into a minimum size, which is 37 m2. In addition, the housing estate also has parking spaces (Figure 1), which there is no rule for occupying. According to the residents, there is a lottery for the parking spaces.

Figure 1: Parking spaces on R. Videiras.

Source: Fieldwork, 2015.

According to the same people, there are problems with building in vertical condominiums, because it is not possible to increase the number of residential units. This factor is seen as a negative point in this type of construction. We see this in the words of Resident No. 18 of the Videiras residential complex, when he says

"My family consists of five people. As our apartment only has two bedrooms, our three children have to share the same room. If we had a house, it wouldn't affect us at all, because we could make a little room to accommodate them better. The government also needs to think about families with more members".

You can also see that there is concern about accessibility for people with special needs. In Residencial Videiras, the apartments on the first floor have been designed for people with mobility problems, and there are access ramps to the blocks (Figure 2).

Figure 3: Access ramp to the R. Videiras blocks.

Source: Fieldwork, 2015.

In addition, all the blocks have water tanks (Figure 4). However, many of the interviewees reported that on days when there is a shortage of water, it is not enough to meet the household's demand.

Figure 4: Hydric structure of R. Videiras.

Source: Fieldwork, 2015.

The guidelines for building semi-detached houses also have the same structure for the number of rooms, but the minimum floor area required is 32 m² (Figure 5). It's worth noting that Santa Maria's single-storey houses have patios that allow for the expansion of residences.

Figure 5: Original unmodified house

Source: Fieldwork, 2015

It can be seen that this house has the same built-up area as when it was delivered, just with a protective fence. It also has a significant amount of free space which can be used for a variety of purposes. This benefits the residents, who have the freedom and space to make any changes they deem necessary.

During the fieldwork, it was noticed that around 70% of the houses had undergone some kind of modification, with the addition of more rooms (Figure 6). According to one resident of Zilda Arns, the size of the plots is ideal, and it also allows for the construction of more rooms.

At the house in question, it can be seen that the resident has changed the structure of the house quite a lot. According to him, a covered garage and a kiosk with a barbecue were built. According to the owner, the house currently has approximately 55 m2 of built area. Still according to the resident, the original size of the houses is satisfactory for small families, but for families with a larger number of members, it is necessary to remodel the built area to make them more comfortable.

Figure 6: House with modified floor area.

The houses are delivered ready, with all the finishing touches, including electrical wiring and tiled floors. In addition, all the houses have solar panels (Figure 7), which are used to heat the water for showers. This shows the concern for using clean energies, making this a project with a high degree of sustainability.

Another point surveyed was current housing conditions. A total of 74 questionnaires were carried out in all the residences. Of these, only 22, or approximately 30%, reported problems with the structure of the house. Of these, 55% reported that the houses had cracks, 30% said that the walls were leaking and 15% reported that the houses were very damp, which favors the proliferation of fungi that are harmful to human health.

Figura 7:   House with solar heating

It can therefore be seen that, overall, the houses are considered satisfactory by their residents. In addition, the quality of the materials used in the construction was of good quality, as no major structural problems were reported.

4.5.2 Social areas: leisure and recreation

One of the main uses of modern housing is to provide well-being and comfort for human beings. In order to do this, large construction companies are sparing no expense. Affordable housing is not far from this, and with the aim of improving the well-being of the people living there, numerous leisure and recreational areas have been built.

During the fieldwork, the facilities and conditions in these areas were analyzed. Social areas included squares, parks, kiosks and sports courts.

Based on the on-site study, it can be seen that all the residences have the following facilities: playground, sports court, kiosk (or party room) with barbecue, as well as green areas with benches and other amenities.

Firstly, during the interviews, residents were asked what they thought of the leisure areas. They all said there were some, but they weren't used because they were degraded. This fact was noted during the observations. Of the three residences studied, only Videiras has its leisure areas in good condition, precisely because it is a closed condominium. The others have been severely damaged and are unfit for use. Residencial Zilda Arns is in the worst condition.

Residencial Videiras has the following social areas: closed party room (Figure 8), with barbecue, tables, chairs and kitchen utensils.

Figura 8: Party room at R. Videiras.

Source: Fieldwork, 2015.

In addition, there is a multi-sports court (Figure 9) which the residents themselves are responsible for maintaining and looking after. It is important to highlight the importance of this court, as it makes it possible to set up social projects aimed at encouraging sport, which could be an important social issue for the reality of the young people who live there.

Figura 9: Multi-sports court R. Videiras.

There is also a children's square for the recreation of its residents, where there is a wide variety of toys for younger children (Figure 10). The residents are also responsible for maintaining and cleaning the area.

It's important to note that these leisure areas at Residencial Videiras are in a good state of repair, but it's clear that they need some work. We noticed some graffiti on the walls of the party room, as we can see in the photos below.

Figura 10: Children's Square R. Videiras

The Zilda Arns Residence, unlike the previous one, does not have a closed party room, but it does have an open kiosk with a barbecue, but it is clear that the place is unfit for use due to vandalism (Figure 11).

It is important to highlight this fact, as it is clear that this residential area is the one with the lowest level of education and average income, which often reflects the lack of awareness on the part of residents to preserve these places. In addition, a large number of criminal gangs are currently using this location, as it is furthest from the center of the municipality and is a low-traffic area, for the consumption and sale of drugs, among other crimes.

This increase in the number of police incidents involving residents and non-residents of the housing estate has worried those who live in the area, as well as being a concern for the governing bodies too. The government and the police have increased extensive policing in the area to curb the actions of criminals.

Figura 11: Kiosk on R. Zilda Arns.

Source: Fieldwork, 2015.

Still analyzing Zilda Arns' leisure areas, it was noted that the housing estate also has a multi-sports court (Figure 12). According to the residents, the community is responsible for the maintenance and cleaning of the area, and joint efforts are made to cut the grass, among other measures.

There is also a children's square inside the housing estate, which is intended for children (Figure 13). However, it is important to note that it is in a state of complete abandonment, with vegetation and garbage taking over the place. Cleaning and maintenance is also the responsibility of the residents, but they have no interest in tidying it up.

Figure 12: R. Zilda Arns sports court.

Source: Fieldwork, 2015.

Figure 13: Children's square R. Zilda Arns.

Source: Fieldwork, 2015.

Another point that caught the eye during the interviews was that residents pointed out that it is the residents themselves who damage the environment, and said that vigilance is needed in these places, especially at night. Many of the residents say that they stop going to these leisure areas out of fear, due to the many cases of crime in the area.

Residencial Dom Ivo Lorscheiter also has an open kiosk with a barbecue (Figure 14). It is in a good state of repair, although there is some graffiti inside. The maintenance and cleaning of the kiosk is the responsibility of the residents' association, which has provided for a fine for residents caught damaging the area.

Figure 14: Kiosk on Dom Ivo Lorscheiter Street.

Source: Fieldwork, 2015.

According to the residents, the kiosk was recently renovated with money from the residents themselves. They also reported that during the night, the place ends up becoming a place for illegal acts, such as drug and alcohol consumption. The residents also stressed the need for specialized surveillance in the residential area at night.

In addition to the kiosk, in keeping with the pattern of the other housing estates, there is also a children's square (Figure 15). It's worth noting that this is the one in the worst state of repair, where most of the toys have already been destroyed and stolen.

Figure 15: Children's square on Dom Ivo Lorscheiter Street.

Source: Fieldwork, 2015.

In general, the popular housing estates in Santa Maria have leisure and recreation areas, but they are highly degraded. However, according to residents' reports, it is the residents themselves who vandalize the areas, thus creating numerous problems for the entire population.

It is important that public managers and residents are aware of the importance of these recreational areas for the development of quality of life. Governments should also be asked to take measures to ensure safety in these areas.

4.6 Health, education, safety and transportation

There are some basic rights for human dignity, including health and education. Safety and quality public transportation are also necessary. With this in mind, a survey was carried out on these factors in the residential areas under study. The collection of this data was also based on the questionnaires applied during the fieldwork, and the reality experienced on site.

As already mentioned, Residencial Videiras differs from the others in that it is located in a neighborhood close to the center and benefits from basic services such as security and education, as it is close to barracks and numerous schools. Despite its proximity to barracks, police incidents are quite common in the area, mainly due to quarrels between neighbors. Transport and health are very good. The residential area is close to the Patronato emergency room and has a wide range of public transport. There are bus stops in front of the residential gate, making it easier for residents to get around. It was also noted that garbage is collected three times a week in the housing complex (Figure 17).

The cases of Residencial Zilda Arns and Dom Ivo Lorscheiter are the same. Both have no school or health unit. An Emergency Care Unit (UPA) is in the process of being set up (Figure 18) which will serve the two housing estates plus the residents of Residencial Leonel Brizola, which will be delivered in 2016.

The residents of these housing estates need to travel to Bairro Sâo José to access a health unit, and to do so they have to cross the RSC-287, which is characterized by numerous accidents, putting the lives of those in need of medical attention at risk.

Figura 16: Entrance to R. Videiras.

Source: Fieldwork, 2015.

Figura 17: Garbage collection in R. Videiras

Source: Fieldwork, 2015.

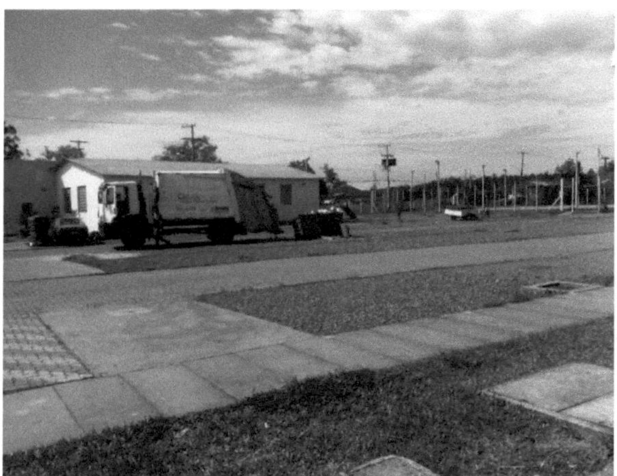

Another negative point is the lack of schools in the housing estates. Young people are forced to study in Vila Maringà or Bairro Sâo José, and in some cases they study in more distant neighborhoods like Cohab Tancredo Neves because there are no vacancies in the nearby schools. According to the residents, an elementary and high school would solve many problems and would be an extra reason for the students to continue studying, as they would be close to home.

Figure 18: Emergency Care Unit on R Dom Ivo Lorscheiter.

Source: Fieldwork, 2015.

Security and transportation are another problem. There is no military post in the area, and there are countless crimes such as muggings, fights and in some cases murders. According to one resident, there is a lot of drug dealing and drinking in the streets at night. As for public transport, residents complained about the reduction in the number of lines that run. There are now few schedules that come into the area. In addition, the lines only pass through the main streets, making residents have to travel several blocks.

The Zilda Arns housing estate has only one covered bus stop (Figure 19), which means that not all residents are able to take shelter there on rainy days. Dom Ivo Lorscheiter has only two covered bus stops, but they are far apart.

Figura 19: Zilda Arns bus stop.

One point that stands out is the state of repair of the bus stops, which are graffitied and in some cases damaged (Figure 20). This shows a lack of commitment on the part of the users themselves.

It can be seen that this destruction of public property by the users themselves shows the lack of social actions that demonstrate the importance of maintaining this infrastructure. According to some residents, there are no campaigns or even community festivities that could integrate the agents that shape the space of the housing estates, and this is seen as a negative point

Figura 20: Don Ivo Lorscheiter Street bus stop damaged

FINAL CONSIDERATIONS

The research showed that the municipality of Santa Aria had two phases of urban and economic structure. The first, between the beginning of the 20th century and the mid-1980s, was focused on the railroad, which brought the town to national prominence because it was located in the central part of the state of Rio Grande do Sul and where most of the passenger and freight trains had to pass. In addition, it was a major factor in the development of the town's urban area, such as the establishment of Vila Belga.

The second stage began with the establishment of the Federal University of Santa Maria in the 1960s, transforming the city into an important university center in the interior of the state. It can be said that this stage was the one that contributed to the increase in real estate pressure in the town, due to the demand that arose for residential units from students arriving in the town.

In addition, the civil service sector, driven not only by the university, but also by the number of barracks set up in the city, especially during the military dictatorship, intensified the process of real estate expansion in the municipality, increasing the number of gated communities.

This real estate pressure has segregated low-income populations into areas unsuitable for housing, thus generating enormous social inequality between Santa Maria's neighborhoods.

As a result, the implementation of the policies studied in this paper has been essential for improving the quality of life of a significant part of the population of Santamari. It is also clear that many of the beneficiaries of this program would not be able to afford to buy their own home, as they do not have the income for the large real estate developments that are currently underway in the municipality.

It is worth emphasizing that the research was successful, as it was possible to achieve the proposed objectives. It was also possible to produce materials that could be used by the city council, such as location maps.

Another positive aspect of carrying out the research was the possibility of being immersed in the reality of the residents. During the days of fieldwork, it was possible to understand the places where each individual lived and what life was like before they were included in the program, and what changes occurred with the benefits they received.

It was also noted that many residents do not understand the importance of contemplation and do not preserve the site. This was particularly evident in the leisure areas, which should be used for moments of relaxation and to provide greater well-being for the residents, but which are abandoned and plundered by the residents themselves.

While the survey was being carried out, a number of facts caught the eye, especially the residents' interest in contributing to it. It was clear that the inhabitants of the housing complex saw it as a way for the reality in which they live to be seen by more people, and also saw it as a way to voice their complaints and praise for the program.

It is worth noting that by carrying out this work, it is possible to forward it to the responsible bodies, so that there can be an analysis of the results obtained and evaluate the improvement of the sites, benefiting the population even more.

There is also a need for improvements in terms of health, education and security, since the sites are located in areas that are not so prime in the urban area of Santa Maria and suffer from crime and a lack of schools and health units. In addition, it is also necessary for the Contemplating bodies to monitor the state of conservation in which each resident keeps their home and the common areas of the residences, demanding that they be preserved and making their residents aware that the recreational and leisure areas belong to them, and that in a good state of conservation they can provide great benefits for them.

In this sense, at the end of the research, it is emphasized that the public housing policies that have been implemented in recent years are of paramount importance for the social development of a country, but it is still necessary for governments to dedicate more resources and continue investing in popular housing developments, thus increasingly improving the reality of the Brazilian low-income population and thus making a better country with less social segregation.

REFERENCES

ABIKO, A. K. **Introduction to housing management**. Sâo Paulo, EPUSP, 1995.Text

technician at the USP Polytechnic School, Department of Civil Construction Engineering, TT/PCC/12.

BATISTA, Ricardo Lopes. Closed popular residential spaces: definition and characterization. In: Brazilian Congress of Geographers, 7. 2014, Vitória. **Electronic Proceedings**... Vitória: UFES, 2014. Available at: <http://www.cbg2014.agb.org.br/resources/anais/1/1403725823_ARQUIVO_ Artigo_final.pdf> . Accessed on: June 14, 2015.

FEDERAL SAVINGS BANK. **Union programs:** My House My

Life. Available at: <http://www.caixa.gov.br/poder-publico/programas-uniao/habitacao/minha-casa-minha-vida/Paginas/default.aspx> Accessed on: May 5, 2015.

CORRÊA, Roberto Lobato. **O Espaço Urbano**. Sâo Paulo: Atica, 2004 (Principios Series).

FEDOZZI, Luciano. **Participatory Budgeting**: reflections on the Porto Alegre experience. Porto Alegre: Tomo, 1999.

JOAO PINHEIRO FOUNDATION. **The Housing Deficit in Brazil**. Belo Horizonte: Joâo Pinheiro Foundation, 1995.

FUNDAÇÂO JOÂO PINHEIRO, Belo Horizonte. **Housing deficit in Brazil 2000.** Belo Horizonte: FJP, 2001. 203 p.

BRAZILIAN INSTITUTE OF GEOGRAPHY AND STATISTICS **(IBGE).**

Available at: <http://www.ibge.gov.br/home/>. Accessed on: October 5, 2015.

IBGE. Synopsis of the 2010 Demographic Census, Rio Grande do Sul. Available at:

<www.censo2010.ibge.gov.br/sinopse/index.php?dados=21&uf=43>.
Accessed on: October 5, 2015.

JUNQUEIRA, Anna Cecilia; VITA, Marcos. **The desires of the middle class**.
Veja. sâo Paulo : Abril, ed. 1739, ano 35, n. 7, p. 98-105, feb. 20, 2002.

LARCHER, José Valter Monteiro. **Guidelines for improving design and construction solutions in the expansion of social housing.** 2005. 189 f. Dissertation (Master's Degree in Civil Engineering) - Federal University of Paranà, 2005.

Ministry of Cities (Brazil). National Housing Secretariat. **Municipal housing deficit in Brazil 2010**. Belo Horizonte. Nov. 2013. Available at: < http://www.fjp.mg.gov.br/index.php/docman/cei/deficit-habitacional/216-deficit-habitacional-municipal-no-brasil-2010/file>. Accessed on: April 14, 2015.

PILAR, Adriana Medianeira Rodrigues. **Irregular occupations along the BR 287 highway in Santa Maria/RS**. 2009. 125 f. Dissertation (Master's in Geography) - Federal University of santa Maria, 2009.

Santa Maria City Hall - Santa Maria **Development Agency:** Santa Maria in data. Available at: < http://santamariaemdados.com.br/1-aspectos-gerais/1-3-historia-do- municipio/> Accessed on: Sept. 21, 2015.

Santa Maria City Hall - **City Office**. Available at: <http://www.escritoriodacidade.net.br>. Accessed on: September 21, 2015

Santa Maria City Hall - Municipal **Department** of **Social Development.** Available at:

<https://www.santamaria.rs.gov.br/secao/processo_seletivo>. Accessed on: September 21, 2015.

Santa Maria City Hall - **Municipal Department** of **Housing and Land Regularization.** Available at: <http://www.santamaria.rs.gov.br/habitacao/>.

Accessed on: September 21, 2015.

Santa Maria City Hall - **Department of Communication and Institutional Programming.** Available at:

<http://www.santamaria.rs.gov.br/noticias/8441-residencial-leonel-brizola-assegurados-r-27-milhoes-para-mais-362-casas>. Accessed on: September 21, 2015.

RAPOPORT, Anatol. **Cultural origins of architecture**. In Introduction to architecture. Rio de Janeiro: Editora Campus, 1984.

REIS, A. T. L.. **Original and modified social housing: spatial configurations and residents' attitudes**. In: IX ENTAC - National Meeting of Built Environment Technology - Cooperation and Social Responsibility, 2002, Foz do Iguaçu. ENTAC - 1993 to 2002 - First Collection of Proceedings of National Meetings on Built Environment Technology. Foz do Iguaçu: ANTAC, 2002. v. 1.

RODRIGUES, Arlete Moysés. **Housing in Brazilian cities**. 7 ed. Sâo Paulo: Contexto, 1997. (Coleçâo repensando a geografia)

SANTOS, C. H. dos. **Federal Housing Policies in Brazil: 1964/1998**. Institute of Applied Economic Research. Brasilia: IPEA, 1999.

SANTOS, C. N. F. V. **Novidades nos modos de Urbanização Brasileiros**. In Habitaçâo em questâo. 2 ed. Rio de Janeiro: Zahar editores, 1981. 196p

SUBSECRETARIAT OF STATE FOR PUBLIC AND SOCIAL ORDER OF THE FEDERAL DISTRICT. **Area de Risco**. Brasilia, 2015. Available at: <http://www.seops.df.gov.br/frentes-de-fiscalizacao/2012-08-21-17-01-06/area-de-risco.html>. Accessed on May 24, 2015.

ANNEXES

ANNEX A: Fieldwork questionnaire

FEDERAL UNIVERSITY OF SANTA MARIA

CENTER FOR NATURAL AND EXACT SCIENCES

GEOSCIENCES DEPARTMENT

FIELDWORK FOR THE DEGREE COURSE

Advisor: Profª . Drª . Sandra Ana Bolfe

Student: Ricardo Stedile Neto

1 - How many people live in the house?

MEMBERS OF RESIDENCE	AGE	EDUCATION	SALARY
FATHER			
MAE			
SON			
SON			
SON			
OTHER			

2 - How did you find out about the Minha Casa Minha Vida Program and what was the application process like?

3 - Where was your old home? Was it risky?

4 - How long have you lived in the housing complex?

5 - HOUSE STRUCTURE

SIZE OF HOUSE	
SIZE OF LAND	
NUMBER OF ROOMS	
GREEN AREA	
SEWAGE	
ENERGY	

6 - Have you made any physical changes to your home? Which? WHY?

7 - Does the house have any structural problems? If so, which ones (cracks, leaks, etc.).

8 - How do you see the residential area in terms of physical and social infrastructure - Streets:

- Homes,

- Leisure areas,

- Transportation,

- Education,

- Health,

9 - What has changed in your life since being considered for the program?

10 - In your opinion, what are the positive and negative points of the residence?

11 - With regard to the negative points, what do you think needs to be improved most urgently?

12 - Do you think that public housing policies are satisfactory or do they still need more investment from the government?

ANNEX B: Lyrics to the song "Santa Maria" by Beto Pires

Santa Maria

Beto Pires

My monument roads and rails

My nostalgia for this time that's gone

This Cerrito

These mountains guard me

And they'll still keep it if one day I come back to you

Holy Mary keep these hills for me

That in its fountains there is the sound of prayer

Santa Maria da Boca do Monte

For you my song, song and song...

Sun in Praça Presidente, warm is your warmth

Lots of bands on the balcony and a blanket over my ear

So many different lives, so many people come and go

Uncertainty for those entering, but nostalgia for those leaving...

Goodbye at the station for those who haven't given it yet

You don't understand who's going who's coming

It's sad to feel the sound of longing when the train whistle goes away

Santa Maria, Maria da Graça, sweet girl, maria fumaça

So green so full of herself

It makes me want to sing to you

Buy your books fast and straightforward online - at one of world's fastest growing online book stores! Environmentally sound due to Print-on-Demand technologies.

Buy your books online at
www.morebooks.shop

Kaufen Sie Ihre Bücher schnell und unkompliziert online – auf einer der am schnellsten wachsenden Buchhandelsplattformen weltweit! Dank Print-On-Demand umwelt- und ressourcenschonend produziert.

Bücher schneller online kaufen
www.morebooks.shop

info@omniscriptum.com
www.omniscriptum.com

Printed by Books on Demand GmbH, Norderstedt / Germany